What Patients Want

Anecdotes and Advice

R. Lynn Barnett

Disclaimer

Events in this book are written from the author's best recollections. This book is not meant to deliver medical advice.

No part of this publication may be reproduced, stored in or introduced into a retrieval system, or transmitted in any form, or by any means (electronic, mechanical, photocopying, recording or otherwise) without the prior written permission of both the copyright owner and the publisher of this book.

Published by R. Lynn Barnett P.O. Box 3762 Alpharetta, Ga. 30023

Dedication

This book is dedicated to doctors' patients and patient doctors, as well as to my husband, whose patience is most appreciated.

Table of Contents

Dr Gralan —   Thanksgiving 2013

I've always said that you were my favorite "back-up," for Dr. Gumer. I know that that makes it sound like you're a back-up singer for a group.

Well, as a "back-up," you have been "Supreme."

Hope you enjoy the book. If you do, tell your friends. If you don't, tell no one. ☺

(It should be available on Amazon within a few months). (It's on lulu.com now).

Rosanne Lynn Barnett
publishing books as:
R. Lynn Barnett

R. Lynn Barnett
PO Box 3762
Alpharetta, GA 30023-3762
inezirv@aol.com

What Patients Want

Almost everyone at some point in their lives will be involved in the medical establishment; either for well-visits, or worse, for sick visits. They will either be defined as patients themselves, or their kids, spouses, friends will be, and they'll have to advocate for them. This book is in no way scientific. Rather, it's a compilation of my observations as to what patients want, from their doctors, their insurance companies, and even themselves.

I have dedicated this book to doctors' patients and patient doctors, because both groups need each other. (They are also not necessarily mutually exclusive). The Mother Goose nursery rhyme that alludes to the way my hair curls these days, is indicative to some people of the state of medical care. "There was a little girl who had a little curl, right in the middle of her forehead. When she was good, she was very, very good, and when she was bad, she was horrid." For some, "when medical care has been good, it has been very,very good, and when it has been bad, it has been horrid."

I have been lucky; my care has been good, albeit, with moments of frustration. Hence, two of the purposes of the book: to vent (and to share). The "Grimm" brothers wrote fairy tales. Medical care should not be a "grim" tale. Sometimes I feel like screaming, "Lab rats unite." There is comfort in shared experiences and in humor. It can be a-"mazing."

To Market, to Market…

The Mother Goose nursery rhyme says, "To market, to market, buy a fat pig." My take on this is, "To market, to market, to sell a thin book." (It should only have been written by a thin author. A few pounds have got to go).

The "germ" of the idea for this book came after our dog died of a recurrent illness. My husband and I were ruminating about his care, and that ruminating was illuminating. I was thinking how blissfully simple the coordination of his care was, and how I hoped some 2-legged friends could say the same about their medical travails. During the process of writing this, a major overhaul of the healthcare industry was in progress. In that respect, the book is timely, but since there will always be doctors and patients, it will also be timeless.

I was asked for whom I'd be marketing this book; who would be my target audience? As I thought about this, I was thinking (hoping) there might be several: an academic audience (medical, nursing and allied health students), practicing physicians, patients and then everyone else.

The students and doctors I'd hope would/could benefit from getting insight into how life is, from the perspective of the person in the white paper gown vs. the white lab coat. This includes drafty rooms and even draftier attire as well as apprehension (of everything: the known, the unknown, and everything in between). Some of my friends have reported a "chilly" attitude as well, if you get my drift, (or my draft, as the case may be). No one wants to be adrift in a draft.

"Everyone else" includes anyone who has dealt with the "medical establishment," including medical offices, health insurance companies, etc. If you are one of the "patient" people dealing with this, know that you are not alone. As with so many aspects of life, that's so important. We have all dealt with bureaucratic frustrations, being put on "hold," for interminable lengths of time, or worse yet, being put on "ignore." Been there, done that. I can hear choruses of "me, too." Let's all sing in unison. A shared sense of anything is good for the soul; a sense of belonging is all part of the human condition.

Speaking about conditions, it seems like medicine has become unconditionally conditional. At one time, if you had a pre-existing condition, an insurance policy might not have covered you. Well, we were all "spanked" at birth; does that "injury" constitute a pre-existing condition? I had heard that a baby was initially denied coverage because he was too fat. What was the skinny on that? I'm happy to say, the insurance company has done a reversal regarding this.

Your doctor might see you on the condition that you have the proper forms, even if you weren't told that you needed that specific paperwork. If you need a referral, and the person in charge of referrals is gone for a week or so, heaven help you. I'd like to scream, on condition that it won't strain my throat, adding to my medical bills.

Some people have asked if I would address their particular group, be it a professional group, book club, commencement address, etc. Yes, I'd love to. I have found that the reactions to some parts of the book are different when they're heard, vs. read. Our new dog laughed when she heard it, anyway. Another book in the making.

When I told a friend about some of my thoughts regarding this book, she said she'd tell people about it and spread the word, but we couldn't think of the word agent. I told her since she'd be promoting my book to her friends, I guess she'd be a "bookie." Since she was new at it, I guess she'd be a "rookie bookie." No "bets" on whether her friends would like it or not.

As with most things in life, luck and timing play a big role in your health. I feel that we have some control over our health though, such as buying healthy food when you "go to market." I hope this book provides "food for thought," whether you wind up in a doctor's office or not.

Speaking of doctors, if any of my doctors are reading this, it has been a pleasure to call you my doctors, and a privilege to call you my friends. Believe me, when you've got a needle, a speculum, a colonoscope and a blow torch coming at you, it's good to be on good terms. You have all been voted "MVP"'s-Most Valuable Physicians .

If you're a patient reading this, I hope this book gives you empathy when you need it, a laugh when you want it, and strength when you think you had none left.

Listen To The Dog, Doctors

When our late, sometimes great dog would sleep, he'd snore. He'd also fake a snore when he didn't want to be bothered by me. Since he knew "I'd let sleeping dogs lie," (in this case, he'd "lie" in more ways than one, the little faker), a snore could keep me away. Boy, could he be a stinker, (there, too, in more ways than one). He didn't take things "lying down," (and neither should we).

If you're a doctor, I don't think any of your patients will snore with boredom, at least not in front of you, but they might indeed tune you out. This is especially true if you're telling them something that they may not fully understand. As a doctor, I think it's up to you to pay attention to that possibility, and to be aware of their responsiveness to you. That responsiveness, or lack thereof, could be telling indeed.

Also, if a patient comes with a spouse, (a useful addition), sometimes the spouse is more likely to facilitate questions regarding potential adverse effects of medicine, post-operative procedures, etc. Animals are great for analogies. The right atmosphere (both physical and psychological) can make a patient come out of his or her shell (like a turtle) and make them take their head out of the sand (like an ostrich).

I must say that when my dog was neutered, the vet asked him (him being my husband, not the dog) if he had any questions. (I think the dog might have raised his paw and said, "I object"). My husband must have asked the vet nineteen questions. As we left, I heard my husband mutter, or should I say "mutt"-er under his breath that "nothing better happen to that dog."

I had an outpatient procedure done around the same time. The doctor said he's fix me right up. (I guess my dog and I were both getting "fixed," so to speak). Not once did I hear my husband (m)utter, "Nothing better happen to that wife."

Ironically, I actually understand this. When our female dog had her first set of puppy vaccines, she whined and wailed, (which meant that I did about the same). We all survived, even though she didn't like having shots in her "tush," (rhymes with push), derriere, whatever "medical" term you prefer. A few days later my husband had his flu shot and he sort of recoiled, winced and said, "Ouch." I'm thinking, "You're a big boy, get over it." With the dog, I was almost apoplectic; with the hubby, not so much. I guess I was more apathetic. Pathetic, isn't it? It's good to be a dog, at least one of ours.

"You Used the Wrong Finger"

When I was learning to type, I had a teacher who told me that for certain letters, I was using the wrong finger, and I should try again, using the correct finger. I was so frustrated. (Boy, could I give some companies the finger, especially when greeted by an automated voice, giving instructions that I sometimes can't hear, and other times can't understand).

My junior high typing frustration was a sentiment ahead of its time. I once called an insurance company, and it had instructions for a variety of languages. I initially missed what you pressed for English (it wasn't as simple as 1) and I had to hear at least 20 other languages before the computer loop went back to English. The generosity of inclusion was superseded by the fact that I just wanted to speak to someone who spoke English. Sprechen sie Deutsch? (Do you speak German)?

Aggravation is not the only universally understood language. Yiddish (which is part Hebrew and part German) often can fall into that category. When I was at the OB/GYN's office, a friend of mine (who isn't Jewish) said the best thing about having a baby was that she wouldn't have to "schlep" to the doctor's office any more, (or at least not as often). I can hear her now at the baby's christening, thanking everyone for schlepping to the church.

Sometimes a patient's question of an insurance company rep does not conveniently fit into neatly assigned categories of "Press 1 for this and 2 for that." I have been known to plead for a representative. Sometimes the computerized voice recognizes the tense tone in my voice and it will say, somewhat soothingly, "We're sorry if we've caused you any undue stress. A representative will be right with you."

Once, after waiting an interminable amount of time for the insurance company rep, my husband came home from work and I had the phone to my ear, still on hold. I told my husband that I felt like I fought all the time with insurance companies. I felt like I was in gym (P.E.) in elementary school, climbing the ropes, holding on for dear life. Now, I'm climbing the walls, still holding on for dear life. At that point, a voice on the other end of the line said, "I didn't understand your response." I was so flustered, I wasn't even sure if it was a computerized voice or a real person. I made some lame excuse and hung up.

Many of us women/girls (whatever term you prefer) were raised to be nice, but just because we're raised to be good girls, it doesn't mean that we should be treated like a dog, or maybe it does, or it should. That will be another "tail"…

We Want to Be Treated like our Dogs

When our dog was sick, (diagnosed with cancer), the vet said, "Bring him in." He didn't say, as some "people doctors" have said, "Well, I have an opening two weeks from this Friday." I know that schedules are set and that other people have appointments, but there must be a better way.

When our late, great pooch, (well, not always great, but his qualities are improving in hindsight) had a run-in with a car, (surprisingly, he gave the car a run for its money), we took him to an emergency vet clinic. The next day, both the emergency vet and our regular vet called to check on him. I know someone who was hospitalized 10 years ago, and she's still waiting for her internist to make a follow-up call. Doggone!

I often feel like a dog when I go to the doctor's office. After all, I "sit," "shake" and "roll over." Sometimes I think they're going to ask me to "play dead," but as long as it's "playing dead," then I think it's OK. Patients don't need to be spoiled pooches, but we don't want to be circus dogs either, jumping through hoops.

It would be most helpful if a patient calls and asks to speak to the doctor, if the receptionist could give a window as to when the doctor might be able to call back.

Some doctors call periodically throughout the day, whereas others tend to call after 5. It would be enormously helpful regarding how a patient spends his/her time. Many people don't want to be tethered to a phone. Of course, most people have cell phones these days, but there are some older people who don't, yet they still have lives outside their houses.

In addition, people can plan to be in a less public place, where they could have a more personal conversation with their doctor, if they had a window of expectation. After all, we don't have "9 lives" in order to wait for a call.

I was once waiting with bated breath for a call back from the doctor. I received a call back, but it was for another patient. They somehow switched my name with that of another patient's. I guess that's a whole new meaning for "bate" and switch.

I have noticed, much to my dismay frankly, that I often get more of a response, and have more action taken, when I act like a @#$%!, (think of a dog) , while I'm put on ignore when I act like a lady! Since when do we reward "bad behavior", and punish good behavior? I got more satisfaction (from people taking my complaints seriously) when I was a little@#$%!.

When our late, (sometimes) great dog would "speak," a lot of times he'd end a tirade of barking with "arf, arf," a two "word," deliberate declaration. I told my husband that maybe it meant, "Thank you," as in, "I've barked continually to let you know that I don't like you other dogs in my cul-de-sac, and I wish you'd leave. Thank you."

Even if it was meant in a mean-spirited way, I convinced myself that that's what he meant, even though I think "arf,arf" really meant "drop _ _ _ _." My husband thought it was somewhere in between, such as, "So there!"

I think us human patients often need to adopt that "so there" mentality. I have known people to play sports after surgery, when they're bruised and bandaged. Women, and especially young girls, can feel self-conscious about a scar, a bandage, etc. I think we all need to adopt the mindset, "Arf, arf, so there." "I'm playing tennis or golf or swimming, etc., with a scar, bandage, whatever. And I still like playing the sport. So there!" "I might have no hair, so there." I'm sure you can think of some other two-word phrases to address any number of frustrating situations. Have at it.

PPO's

We all know that PPO stands for Preferred Provider Organization, but I have my own definitions for the acronym.

Persistence **P**ays **O**ff- Many times I'm sure that insurance companies are bombarded with paperwork. If you don't get a response in the time that your company is required to answer an appeal, make a call. Perhaps they legitimately didn't get your claim, or it got way-laid someplace in the organization. If you think that you have a legitimate complaint, follow through. You know what they say, "Fool me once, shame on you. Fool me twice, shame on me." It's up to you as a patient to sometimes take the initiative.

Politeness **P**ays **O**ff-Regarding the above advice of taking the initiative, it can be done in a polite way. The insurance company rep is just doing his/her job.

Degrading them or their company won't endear you to them. Sometimes situations don't initially fall within the guidelines of a covered procedure. Hence, the idea of an appeal, where sometimes those restrictive lines of what's covered and what's not can be bent a little. But, it doesn't hurt to be nice in the process. Even though you might be understandably frustrated with a situation, try not to take it out on the person at the other end. Try to be a "ppo" yourself- a "patient patient, often."

Patience **Pays Off**- If your insurance company has 30 days, for example, to respond to an appeal, don't ask them after two days. I know it's hard to wait, but patience for the patients is something to strive for, even though it might be difficult.

Perspicacity **Pays Off**- That was one of my father's favorite words. It means to have unusual insight into a situation. Sometimes you might have to delve deeply into a situation to get some results. If the insurance company informs you that they need more medical information, you might have to request that information from all doctors who were involved with your care, maybe not just the one doctor who first comes to mind. Follow up with requests for that additional information, if necessary.

You might have your own definitions of what PPO stands for. I'd give you more examples, but I'm PPO'd-Pretty Pooped Out.

It's Appalling, I mean Appealing

So many people are intimidated by the prospect of an appeal. They think they have to be a combination of idealized fictionalized and/or real doctors and/or lawyers. Not so. If you present documenting evidence, such as lab work, doctor's notes and your own summary of events, you might win your appeal.

Again, even though your frustration of events can be palpable, I have found that summarizing the situation in a concise, succinct manner is often the most effective. I'm not saying to be dispassionate; certainly let your frustration show, but in a succinct manner. Keep copies of documentation and correspondence, so you can refer to them if need be. Read the appeals policy of your particular company, regarding how long you have to file a claim, where to send it, can you fax it, etc. It's time well spent. Remember, even though we might consider the spirit of the law to be more important than the letter of the law, I betcha if an insurance company says that you have x number of days to appeal, they mean x, not x + 1.

I knew of a case where the second appeal letter wasn't signed. The person's name was on it, but he forgot to sign it. Hence, it sat on someone's desk, until the patient inquired.

A friend was cleaning out her closet, and an orthopedic medical appliance fell out. Her husband said, almost wistfully, "That was our first appeal." Some couples wax poetic about wedding pictures falling out of closets. Not them.

My husband, though, can wax romantic, or at least I thought he could. He was on the phone many moons ago, shortly after we had gotten married. I heard him say, "She's appealing." I thought he was talking about me. Well, he was, but not in the way that I thought. I was appealing an insurance claim.

WWW.= What We Want

I was recently looking on the web for a dog to get from a rescue group. Many of their traits are highlighted. Some were good with kids but not other dogs. Others were good with cats but not with kids. You get the idea. I could specify that I wanted the dog to be hypoallergenic and housebroken. After a long search, I just wanted to do a computer search for the following: "dog who will love me for the rest of my life."

I'm often tempted to do something similar when looking for a new doctor, too. (Although I'm glad my doctors are both of the housebroken and non-shedding varieties), I really want to search for a: "Doc who will listen to me for the rest of my life, or as long as he or she is on my plan, whichever comes first." Oh, if it were only so easy. You can often find out about a doctor's credentials, their training, where they went to med school, etc., but that elusive, hard to define and often harder to find "bedside manner" requires a face-to-face meeting, sometimes several.

Sometimes certain situations just inspire me to think, "Give me a break, will ya?" (On second thought, maybe that's not the best term to use with an orthopaedist). I think that doctors and patients alike need to heed that advice. Sometimes a doctor can be up all night delivering a baby. Sometimes a patient can be up all night with a crying baby. Either way, lack of sleep can cause crankiness on either or both sides. If you're a patient, understand that you might have to wait a while in the waiting room.

Conversely, if you're a doctor, understand that a patient might not be so patient after waiting a long time in a waiting room, (and a sometimes cold one at that). Whether delivering a baby, or delivering bad news, or anything in between, I think understanding is needed on everyone's part.

I think as patients, we need to have somewhat of a partnership with our doctors. I think we not only can, but should have a say in our medical care and medical decisions. Sometimes, we have to take matters into our own hands. For example, it has been difficult to find certain vaccines at times. It is up to us to make phone calls, do a little research, etc., to find out who might have a vaccine that we need. The research might not pan out, but it's worth a shot, so to speak. Sometimes we all need a shot in the arm to take matters into our own hands. Then we get to call the shots.

Even if we don't primarily define ourselves as patients, sometimes several visits to a doctor's office in a relatively short period of time can define you as just that. I had seen my Ob/Gyn several times within a short period of time. At night, my husband and I were watching a show on TV that featured some police. I was mostly asleep in the recliner, with my legs resting on the foot rest, knees slightly bent. One of the policemen (I don't remember if this was a fictionalized show or not), said "Spread 'em." Even being half-asleep, my husband said I spread my legs, as if on an exam table. Sometimes the label "patient" seeps and creeps into our subconscious. (When I'm in a doctor's office in that position, I often feel like a wishbone; I wish I weren't there).

The Initial Visit

You will become familiar with all sorts of initials when you're thrown into the medical maelstrom. These include MRI, CAT scan, PET Scan, X-ray, PRN (as necessary), and a CPM machine, technically called a continuous passive motion machine, but some patients call it a constant pain machine. My all time favorite, though, is what you'll want to call the physical therapist, after repetitive and grueling exercises, when he says, "One more time." I promise you you'll want to say, "GTH." I could even text message it.

Speaking of "pet" scans, it's so tempting to want to play fetch with the dog, but you'll find that depending upon your particular injury or illness, bending and/or throwing might be difficult. My dog looked confused as to why I couldn't be his servant/playmate and pick up the ball as usual, but, he got over it. Your pet will, too. It might even bring out a heretofore undemonstrated compassion in him or her. (Some dogs are naturally compassionate. With others, it takes a little time).

I have to add some other initials, "bc" and "ad", with no disrespect intended. My definitions are "before cancer" and "after diagnosis." I have several friends who are cancer survivors, and there is a difference in your perspective on life, "bc" and "ad." BC can also stand for, in my definition, before computers. People who are diagnosed now have a wealth of information available to them at their fingertips, regarding support groups, treatment options, etc. There is a line drawn in the sand, "bc and "ad." Sometimes, that line is a good thing; you don't sweat the small stuff. Sometimes it's bad, in that it's often in the back of your mind. I used to play on an ALTA tennis team (Atlanta Lawn Tennis Association). I have friends who had pulled muscles, torn ligaments, etc. I am now going to form my own tennis A-L-T-A league, "All Ladies with Torn Anything." As anyone over 30 knows, you just don't heal as well as you "age." (I love it when a 29 year old doctor tells me that). It takes a slight movement and one second to tear something, and weeks to recover.

Have a new concept of time. When a doctor says that it will take a few weeks to recover, he might be thinking of his last patient, a young athlete. I was told that Tiger Woods had knee surgery too, and that he recovered quite well. Tiger Woods is such a phenomenal athlete. We're not all "tigers in the woods." Some of us are more like "hippos on the couch."

Give yourself a break if you're not dancing til dawn by the time the "expected" timetable rolls around. Certainly, question your doctor if you're concerned about your recovery, or lack thereof. But, if he/she listens to your concerns and follows up with you, and says that it might take a little more time, give yourself that time. Everybody, and every body, does not work on the same timetable. I think that we are often multi-tasking ourselves to death, and we feel guilty about having "unproductive" time. (Sometimes, I even have trouble single-tasking). When I had an outpatient procedure, my doctor told me to rest for a few weeks, no heavy lifting, no housekeeping, etc. I said, "Are you sure I can't push a light mop in a week or so?" He said, "No, you need to rest." Here, I have not only a doctor's permission, but doctor's orders to put my feet up and take it easy, and I'm fighting this? Am I nuts? Other household members can pick up the slack (and the slacks and the socks…)

ABCDE

A neighbor has 5 children. The first child's name begins with an A, the second, with a B, etc. Accordingly, the last child's name begins with an E. I was telling that to a friend on the phone, when another friend came to the door. The latter woman overheard me say, "Remember, ABCDE." She said, "Isn't that nice, that you're reminding someone of the signs of skin cancer." ("A for Asymmetrical, B for Borders that are irregular, C for Color, D for Diameter being bigger than a pencil eraser, and E for anything on your skin that's Evolving or changing"). (This, thankfully, has been promoted to the public by the American Cancer Society, The American Academy of Dermatology and the Skin Cancer Foundation, among other organizations). (I have also heard some people refer to E as standing for possibly elevated at times).

I had to laugh, because although when I hear ABCDE, I usually do think of skin cancer warnings, in this case I wasn't. (By the way, although I love kids, I think you have to be extremely organized to run a household with many kids. If I followed suit with the ABCDE pattern, my sixth kid would have to be named "Finished").

ABCDE doesn't necessarily mean that you have skin cancer, and not having ABCDE symptoms doesn't mean that you don't have it. A friend from college who became a doctor looked drained and strained at times, but his eyes were trained to spot a problem, so let a doctor's trained eyes give his/her opinion.

ABCDE does bring to mind that I had two ideas for a PSA (Public Service Announcement).

P-S-A------Pretty Savvy Ad

PSA can have many meanings. One is a test to detect the possibility of prostate cancer. Another meaning is a Public Service Announcement. I like to think that my definition, pretty savvy ad, combines the two; making people aware of cancer detection via public service announcements.

I have an idea for an ad that would highlight the ABCDE warning signs of skin cancer. It would show a teacher with a pointer, pointing to things on a blackboard (or a more contemporary white board). She would say to her "class" (the TV audience)-"I'm going to give you a test, but I'm going to give you the answers. Ready?... Skin cancer, particularly melanoma can look like A-Asymmetrical Borders, B-Borders that are irregular, C-Color, dark, but can also be pink, brown, speckled or have different colors, D-diameter larger than a pencil eraser, E-any mole that's Evolving, or changing, OR F, NONE OF THE ABOVE. That's right. Sometimes, skin cancer doesn't fall into the convenient A-B-C-D-E categories. That's why it's so important to visit your dermatologist regularly. He or she may spot a spot that you or I cannot (and hence be 'Johnny on the spot'). So please, visit your dermatologist regularly. It's one of the most important tests you'll ever take."

Just think, if you have skin cancer that doesn't conveniently fall into those specific ABCDE categories, and if yours is an "F", you can have the thrill of telling the doctor to "Get the F out of here."

While I'm on a roll, I have an idea for another ad. I would like to have an ad where the top half of the screen shows a WOMAN's face, (and it must be a woman, for what you'll soon see will be obvious reasons). The voice-over would say, "This is the face of ED, but in this case, it's not what you think. (ED usually stands for erectile dysfunction). In this case, ED stands for Early Detection, one of the best weapons against cancer. So please, visit your doctors regularly, and have whatever tests and screenings they recommend. Remember, this is the face of ED."

I think the beginning of the ad would make people do a double-take. If it made them look twice, maybe it would make them think, and if it made them think, maybe it would make them act.

Stay Sis, I mean stasis

The word "stasis", as I understand it, means a meeting of the minds when it comes to a certain term. For example, the word "cow" to a meat lover might mean lunch, whereas the same term to someone else might signify a holy animal. I think it's so important, particularly in a doctor/patient relationship, to make sure that everyone is on the same page.

If a doctor says, "use this once a day," I think it's important to clarify any further instructions, such as use once a day in the morning, or at night, or with food, etc. I have gotten medicine from the pharmacy with instructions to take it "with food" and "without food." I called the pharmacist to clarify the apparent conflict of information. She said that since the medicine can cause stomach trouble at times, and I'm prone to that, that I'm better off taking it with food. Get to know your pharmacists. You'd be surprised what an integral part of your life they can become.

One of my funny experiences with a misunderstanding involved an X-ray technician. Shortly before I had this X-ray, when I left home, there was a beautiful model on TV. I was thinking that she certainly was a "10", the concept of idealized beauty, indicative of Bo Derek in the movie of that name, written by Blake Edwards. (I'm more like a 7.5, and that's on a good day).

After I had the X-ray and was getting off the table, the technician asked me if I was still a "10." I told him I never really was. He said that my chart indicated that I was. First, I was flattered, and then I was mortified. Do medical charts rate their patients' looks?! Are our looks a determining factor in our treatment?

Well, he was talking about "10" on a pain scale of 1-10. Stasis anyone?

That misunderstanding is right up there with the time that I was under the weather for a while. When I got up, I made some biscuits. As I was putting them in the oven, my husband said, "Don't over do." I thought that was quite considerate, being that he had been living on frozen dinners for a while. Then he said, "Because the last time you made biscuits, they were overdone."

My husband had mentioned reading something about ACOG to my cousin, who is a doctor. My husband was referring to ACOG (Atlanta Committee for the Olympic Games). My cousin was thinking of ACOG as the American College of Obstetrics and Gynecology. Needless to say, their conversation went round and round a bit until we figured out the confusion.

After that, as a patient, I learned to be ACOG myself, "Always Compliant, (but) On Guard" for any misunderstandings.

Some of us slip and fall on our faces. Others of us have slips of the tongue, and still fall on our faces. I was scheduling my routine mammogram, and I told the scheduler, "I'd like to schedule my screaming, I mean screening mammogram." We both had a good laugh about that.

I once had a student who was planning on being a healthcare professional. Since I was a bit rundown, I said that she could be my nurse. She said, "You have no charge." I thought she meant no vim, no vigor, but she meant that there wouldn't be a charge for me. How nice.

There's more. I was nervous about a dental procedure and I asked my husband if he had the insurance card for mental health, and then I corrected myself and said "dental health." I might be an overly concerned patient, but at least I'm good for a laugh.

4 3-Minute eggs

Many people whom I know have said that they have to continue talking to the doctor in the hallway, because their "allotted" time is up, (often about 12 minutes or so). I always think of an egg timer in those circumstances. I know all about overhead for a doctor: rent or mortgage of office space, insurance (health, malpractice, employees' coverage, etc.), utilities and the like. But patients aren't eggs. Well, we can be good eggs or rotten ones, just like doctors, but we're not timed eggs.

I once tutored some children, and one of them asked me if it was easier to have a baby or to lay an egg. I told him that I had laid a bunch of eggs in my time, but he didn't get that. Then his (younger) sister told me that her friend said that she came from an egg.

There was a look of utter disbelief on her face. I said, "Well, technically, she did. Inside women, we have eggs.

And men," and I stopped. I said that her parents could explain things to her later, and she didn't push it. I could hear myself, in my head "page" her doctor father, who was upstairs with his wife, if the little girl had requested more information. I would have said, "Dr. 'Smith.' Paging Dr. 'Smith'. Please come to your living room, unless you're trying to make an egg yourself, in which case, carry on."

Speaking about eggs, when I had a procedure at the Ob/Gyn's office, when I was trying to conceive, I was put on a table after a certain procedure and told to rest there for a few minutes. An "egg" timer was used to help me keep track of the time. How appropriate.

As with most things in life, timing is everything. I had just taken a test earlier that morning that confirmed that I was mid-month in my monthly cycle, an optimal time to try to conceive. I made the necessary work schedule changes and then went to the bank to do some business before heading off to the doctor's office. At that time, this particular bank charged you for deposit slips, although you got two free ones per month. I hadn't remembered if I had used my two or not, so when I inquired if I'd be charged or not, the teller said, "It depends upon your cycle. You're mid-month, aren't you?" Well yes, I was, but I didn't know how she knew. She said that it was on her computer screen. I had heard of George Orwell's concept of "Big Brother" watching us, from his book "1984," (published in 1949), but this was ridiculous.

She tilted the computer screen for me to see. I don't know what I expected to see; perhaps, "Mid-month? If you're lucky, you'll need diapers in about nine months; they're in aisle six of your local store." Or maybe, for us "girls" who are reading this, (men, you can turn away for a minute), it might have said, "If you're unlucky, there are other items that you'll be needing sooner).

The teller again said, louder, that I wouldn't be charged, and I said, "Sh." She reiterated that since I was mid-month, I'd get the free deposit slips. I was thinking, well, what do I get if I have a baby, free checking? Well, she was talking about my mid-month "billing" cycle. Whew, that was a relief.

Sometimes there are going to be misunderstandings in life. You can "bank" on it.

"Weight" a Minute

Although for practicality sake, a certain amount of time has to be allotted to each patient, there has to be a little wiggle room with that. If the first patient of the day gets bad news, and each patient of that particular day also receives bad news, and as such may need a little more time than the originally allotted time frame, the wait for both the doctor and the patient will multiply exponentially. It's just an inherent part of medicine.

Once I had a doctor say, "I'm sorry about the weight." I said, "You know, I eat 1200 calories a day, I exercise 45 minutes a day. You tell me what else you want me to do!" Well, he blanched, and said, "The wait. W-a-i-t. I kept you waiting about an hour. It's inexcusable. I'm sorry about the wait." Here I was, ducking an insult, and he was apologizing. (Talk about stasis).

Speaking about waiting, I was once asked to wait in a waiting room to await the results of a relative's procedure. There were two waiting rooms in that particular area of the hospital and I asked the receptionist if it mattered which one I waited in, and she said "no."

Well, after quite some time, I was asked to go back and speak with the nurse. She told me that the doctor had gone to the waiting room three times and that I hadn't been there. I told her that I was there, that I hadn't left. She repeated that I wasn't there when the doctor went there. I repeated that I was. Round and round we went. Well, the doctor called me at night to explain the results of my relative's test. I relayed my conversation of earlier in the day, and he said that he had looked in the other waiting room. Talk about "relative" aggravation.

I don't mind a wait in the doctor's office, (except in a cold exam room). I know what it's like to get unexpected news, and need more than the originally anticipated amount of time. To me, if you wait, it means the doctor is taking his/her time with his/her patients. Certainly, scheduling might have to be adjusted if patients are constantly booked back to back with no breathing room for an extra needed few minutes here or there, but I view a wait as a doctor giving a previous patient some needed time. I also might need a little more time. Even recipes can be adjusted. If you can do that for meatloaf, you can do it for me.

Some doctors apologize for a wait, look you in the eye, and give you the time that you need, despite financial pressure to usher you in and out the door. There is no EOB (Explanation of Benefit) needed for that. That's priceless.

B.S.   Bachelor of Science

I know, I know. You thought B.S. stood for something else. Sometimes it does, especially dealing with some rules and regulations, such as when the scheduler of a doctor's appointments is out of town and no one picks up her messages. Consequently, when she returns and then returns your phone call, it is only to tell you that now it's too late to make an appointment because the time allotted by your insurance plan, for the number of visits that you are entitled to, has elapsed. Bachelor of Science!

However, for the purposes of this chapter, B.S. really does stand for Bachelor of Science in the truest sense. Medicine, like any other profession, is a combination of an art and a science. I guess you could call it a Master of Arts and a Master of Science.

A teacher in training can be taught how to teach children to read for example, but it's often an instinctive sense of what approach would work better with which kid. Some kids will learn better through an auditory approach, some with visual, others with kinesthetic. Sometimes it's trial and error until you find what works best for which kid. Sometimes you might need to throw away the book, (although that might be a poor choice of words for a reading teacher).

It's the same with medicine. Cardiology professors, for example, can teach you how to detect cardiac problems, but they can't teach you how to practice medicine with heart.

I was once in an academic setting, and someone who was getting a degree in Counseling was rude and abrupt to someone else who was anxious about a certain situation. And this person was getting a degree in Counseling?! I guess you can be taught listening skills, it doesn't mean that you know how or when to use them. You can lead a horse to water…

I have had many doctors through the years say, as was previously mentioned, "Sometimes you have to throw away the book." Now I'm not suggesting using a rusty needle, even though "the book" says to use a clean one, but I am suggesting that sometimes a doctor needs to listen to his or her gut. I think some of the best medical decisions are made by a combination of not only the cognitive skills that are taught in med school, but by a doctor's gut and heart. (I have been told that Med School deals with your mind in more ways than one). I have been lucky enough to have doctors who have combined all three of the above mentioned traits, and unlucky enough to have doctors who were lucky to have one.

Three Little Words

When you're young and impressionable, the three little words you want to hear are "I love you." Then, you get married, and have a family, and the three words that you value most are "No ironing needed." Then, you wind up being in doctors' offices a lot. There are other sets of three words, that the more I'm in doctors' offices, the more I value them when the doctors relay them:

I'm not sure.

Let me see.

Let's re-visit this.

It is benign.

You are lucky.

I will ask.

Let me investigate.

You are right.

Point well taken.

Trust your instincts.

I believe you.

Have some faith. (Which I interpret as faith within yourself, your doctor, and/or a higher power).

And the two sets of words that I have really come to value and respect: "I don't know" and "I am sorry."

One more thing. Sometimes a doctor can say any of the above without saying a word. Sometimes a look says it all, and that silent look can be more eloquent than any soliloquy.

" R"idiculous

It used to be said that you should eat certain seafood only during months with an "r" in it. I think it had to do with water temperature, and therefore the safety of certain shellfish. When similar reasoning applies to refunds though, it can be ridiculous.

We were owed a refund from a doctor's office. A few months went by, with still no refund. When I called to inquire, I was told that refunds are usually just issued once a month, and if they didn't get to it that particular month, they would do it the following month. The following month, December, comes and goes. Towards the end of the month, I was told that refunds are usually issued at the beginning of the month, but that month it didn't hold true, and then it was holiday time. "Hell(o)"?!

The following month, the person in charge of the refunds had a family situation to deal with. I can empathize, but I'd like to try this with my mortgage company. (Not that I couldn't pay, but I'd have a litany of excuses as to why payment would be delayed).

The month after that I went in person, and still had to wait for the doctor to OK this refund. He apparently had to sign the check. His signature was barely a scribble. (I'm not going to make a crack about doctors' handwriting). If this medical practice practices this kind of treatment every time a refund is due, imagine how much interest they're getting. You do the math. I would, but I'd probably get an ulcer, get treatment, be due a refund, and the whole cycle would start again.

Trust Your Gut

My mother, a cancer survivor, had a pulling and tugging sensation on the left side of her stomach. She told her GI guy, (gastroenterologist, hereby known as the GI guy) who ordered a colonoscopy, even though he initially thought there wasn't a problem. Well, she apparently had often hard-to-detect flat polyps on the right, ascending side of her stomach. Let me tell you, that ascending colon problem was almost ass-ending, (excuse my French). It is a good thing that my mother trusted her gut, in more ways than one. Even though the cancer was on the right, the discomfort was on the left. Go figure.

Some people call it woman's intuition. You don't have to be a woman to have woman's intuition. I think we've all done this; we've picked one restaurant over another, one road over another, one social engagement over another, for no particular reason other than it just felt right.

I think many decisions are made by that "je ne sais quoi" (I don't know what) reasoning.

In medicine, particularly, I think it's important to pay attention to those signs, physical and mental. Sometimes going to one doctor seems like a better choice than another, even if they're both random selections. I'm not saying that you have to stick with your instincts if they're wrong, but I am saying that it doesn't hurt to pay homage to them.

My mother was worried about a little spot on her back, and she wanted it biopsied. Her dermatologist agreed. It was no skin off his back, (although it was skin off of hers). Luckily, it was nothing, but she had a strong enough relationship with her doctor to be able to request a second look.

Speaking of intuition, there's another "in tuition" to consider, the tuition of medical school. Although intellectually a med student knows that he or she will make more than enough money to pay back their student loans if they have them, it's another thing to be socked with that kind of loan in your mid-twenties. When patients see the fruits of a doctor's labor, (in terms of a big house, nice car, etc.), they need to remember that it was sometimes a hard-fought battle to get where they are.

A medical education, like most things in life, has a cycle. You wait for an acceptance letter to the undergraduate college of your choice. As a freshman, you might be a little unsure of yourself, and it's all you can do to find your way around campus. By the time you're a senior, you're more confident and capable, but now, you're waiting for an acceptance letter to med school. Again, during your first year, you might re-visit the uncertainties of a few years back. By the time you're a senior, you're again more confident and competent. You'll have MD after your name.

The internship year will keep you grounded, because you're so exhausted, you can't spend much time patting yourself on the back regarding your new title. After your internship, you're a resident, but you have more senior residents to answer to. Finally, the "training wheels" come off your bi "cycle," and you are no longer in "training." You might join an established practice, where you'll (again) be the low man, (or woman) on the totem pole. After a few years, though, you'll no longer be the new kid on the block, and you'll have no one to answer to except yourself and your patients. By this time though, you might be too burnt out to enjoy it.

If you're a doctor reading this, I hope you don't lose the nerve, verve, passion and compassion you felt when you first graduated.

When I graduated as an undergrad, the entire university had one joint ceremony, and then the respective colleges had their own ceremonies. The excitement and anticipation of the med students was almost palpable. It was contagious, in a good way. (I hope that sense of excitement stays that way). I know it's easy to get jaded, but as doctors, if you can retain a little of that initial inertia and excitement, you'll be rewarded in more ways than one.

If you're a med student reading this, at your graduation, take a minute to think about the shoulders that you have stood upon to get you where you are. These would include not just those of your family and friends, but perhaps the first grade teacher who taught you how to read, so you could read the big medical text books, or the piano teacher who helped develop good eye hand coordination, also a good trait for doctors.

For me, the most important aspect of being a doctor, is to remember that "what" you treat isn't as important as "whom" you treat, and "how" you treat them. That is the most salient tenet of medicine, in my opinion.

As a med student, and then later as a practicing physician, you'll learn to be a soldier, willing to go to battle for your patients. You won't be a "soldier of fortune," so to speak, but if you make a fortune in the process, as long as that's not your sole goal, that's fine, (but it should be your "soul" goal).

Soldier on.

MD

We think of MD as standing for Medical Doctor, but for me, it can stand for so much more. I once tutored some kids whose parents were doctors. Once, one of the kids was really stressed out. I told her to meditate. (I inadvertently, initially wrote medicate, but I indeed told her to meditate), to mediate the situation. She tried meditating and said that it helped. The next time I saw her, she mentioned in passing that when she and her family went out for dinner, she mixed lemonade and lemon-lime soda, my favorite drink when I'd go to her house. I told her she'd be an MD either way, either Medical Doctor, or Meditator, Drinker. (Or maybe both).

I must say that in this particular case, the kid's going to make a wonderful MD, in the truest sense of the word. One night, I was tutoring her, and my eyes were straining to see the somewhat small print. She unceremoniously switched books with me, because her book had larger print. No fanfare, no nothing, she just did it. You can't teach that. I told her that she had an instinctive sense of what to do, and she'd be a good doctor because of that. She said that her parents thought that she'd be a good doctor because of her grades. Both are important. I can sing her praises.

Maybe MD can stand for Medical Dictionary. I tutored another child who was playing a word game with me. He used the word parthenogenesis. He told me that it meant "reproduction of fertilizer." That didn't make sense to me, so I looked at his notes, which may or may not have come from a medical dictionary, and his notes said, "reproduction, without fertilization," (like what you can do with a tomato plant). I started to explain the difference and then I thought that by the time he knew how to fertilize a lawn, he'd know how to fertilize other things.

I also think MD can stand for Major Decisions. We have to face those all the time. Do we use treatment A or Treatment B? Plan A or Plan B? Dr. A or Dr. B? Insurance A or Insurance B?

It can certainly stand for Minor Disappointment. I had some outpatient procedures done, which included somewhat painful shots and a punch biopsy, (I guess you can say that I was "shot and punched,") and I looked it. I was red and swollen. I couldn't wash my hair for a day due to the biopsy. The doctor left me in stitches, but I wasn't laughing. Small price to pay though, to rule out a problem. However, that night I get a call from my boss telling me that a photographer is coming the next day to take pictures of us! Do you believe this?! I have come to realize, it happens.

I must add that MD does not, and should not stand for Medical Deity. Doctors, after all, are just people, but, it's amazing that some people don't recognize that.

I have a friend who is a doctor, named John. I have another friend who is intimidated by doctors. I told her that although doctors are highly intelligent and well-educated people, at the end of the day, (here's where I put my foot in my mouth) we're all just "Johns."

I knew a doctor who lived in a humid climate, and his house was infiltrated by termites. I mentioned this to a friend, who couldn't believe that a doctor was saddled with such aggravation. What are the termites supposed to do, say, "OK fellas, let's not bother the doc, let's hit the accountant next door." Don't get me wrong; I wish all of my doctors health and wealth, in that order, but it doesn't mean that a termite might not visit them every now and then.

Many med students over the years have told me that sometimes they feel like MD stands for "Mildly Deranged," "Manically Depressed," " Mainly Dead-tired," or "Mostly Depraved and/or Deprived." I think at some times, many of us have shared some or all of those feelings. Hopefully, we can change that, or transition to "Mostly Determined." If you're a patient, you need to be determined to fight a disease, the system, (if necessary), etc. If you're a med student or doctor, you have your own systems to fight (med school, insurance companies, etc.). I think it can also stand for "Make a Difference." That can apply to the patient as well as the doctor. The doctor can make a difference in your life by treating a disease, illness, etc. The patient can make a difference by taking a stand, if something doesn't seem right in his/her care. A doctor might know 99% of what happens regarding an illness, but you know 100% about you.

Doctors go through a lot in order to have the privilege of putting MD after their names. They spend long hours studying, refining and sometimes re-defining their craft. They take "gross anatomy" classes. Although I know the technical term of "gross" anatomy, I will tell you that sometimes, gross means, well, gross. And they have to do it anyway.

MD for me can also stand for make-do, but it should not refer to medical care. I think it can refer to making do with older magazines in waiting rooms, paper gowns that are way-too thin, and a longer than expected wait for the doctor. It should not be used in the context of "making do" with sub-standard care.

MD for me can also stand for Minor Disease but with Major Distractions. Even if you have a so-called minor problem, like a head cold, it can still be difficult to do daily tasks. A sore throat and runny nose can make eating, sleeping , and even bending down to tie a shoe seem like daunting feats. If the only medicine that works for your symptoms makes you drowsy, and you have to drive, there really is no choice but to put up with the minor discomfort. It's hard to accept that things have to get done, even if you're not at your best.   I bet you cut other people some slack when they're not up to par. I think we need to learn to do that with ourselves.

Sometimes we feel like turkeys, some of us more than others. I once dropped a bag of groceries in front of a grocery store around Thanksgiving time. I had to replace all of the damaged goods. A doctor friend was surprised that the store didn't offer me a turkey or something. (You'll soon see why I didn't need any more turkey). I blamed myself for not seeing the uneven pavement.

Sometimes though, you can do everything right, and things still go wrong.

I was having an outpatient procedure and I asked the doctor many questions about pain management, recuperation time, etc. The one thing I didn't ask him about was food.

I was determined to be in the best possible shape before-hand. I ate a healthy dinner the night before the procedure: turkey, some fruit, and a cup of tea. At night, my husband and I were watching the news and the reporter announced a major food recall----of turkey! (After some investigation, I found out that my turkey was not part of the recall).

A few days later I had called a friend who was having her own health issues and she told me that she was in the process of having her dressing changed (for a wound). I told her we could call ourselves "turkey and dressing."

I have to end on a funny note. I said something to a friend's child, who responded by saying, "I didn't know that, Dude." The first time I didn't say anything, but the second time I said that I wasn't a dude, I wasn't even a dudette. I could see my new moniker now, Mrs. Dudette, MD for short. Better yet, maybe just refer to me as "Dudette Barnett."

Sip Your Teas, Cross Your Eyes

I know, the original advice is, "Cross your T's, dot your I's," advising you to pay attention to details, but sometimes, you feel like crossing your eyes in disgust, and drinking your teas for nourishment, when dealing with the specificities and minutiae of medical care.

I know of a case where the extension of use for a specific drug and the specific dosage was approved, and then denied. No one could tell the patient's family what caused the "reversal of fortune," so to speak. The doctor's office verified sending the information to the pharmaceutical company, only to be told by the pharmacy that they never received this information. The patient's family had confirmation numbers. This, was to no avail. During the first phone call, the family member was told that there was no supervisor on duty. During a follow-up phone call minutes later, poof, the magic supervisor appeared. Are your eyes crossed yet?

Apparently, the request had to be made on two separate pieces of paper; one for the dosage of the drug and one for the duration. Both pieces of this information were sent on one piece of paper initially, although the family member, properly, received two confirmation numbers. Eventually, things were straightened out, but I think that cup of tea should have something in it by now. Scotch and tea, anyone?

Once a claim of mine inadvertently had my husband's name on it. It wouldn't be such a big deal, except it was a procedure that is usually performed on a woman. Again, it was rectified. I was getting a little "t"-eed off.

I have to give some advice. "Watch your P's and Q's," as in, if you "pee" in a cup (excuse my French), and the doctor never calls you with the results, that's your "cue" to call and inquire. In my case, the sample was lost (at a most inconvenient time, since my mother was in the hospital at this time). It all worked out in the end (pun intended), but it was a pain in the _ _ _ (also, pun intended).

I was told that misplacing samples is a rare occurrence. R-a-r-e is not a four-letter word. (Well, I guess it is). I must digress for a second. I once brought cookies to a child whom I was tutoring years ago. His mother asked if I had baked them. I said, "No, 'bake' to me is a four-letter word," to which the kid replied, "She's such a good speller."

I have to add one other funny event about tutoring. A student and I were reading a passage about sunscreen use, and of all of the passages that we read, I really stressed getting the questions to that passage right. Since I'm fair-skinned, I'm adamant about sunscreen use. I figure this student might not know how to read, write, or spell, but he would know how to use sunscreen.

Well, back to the matter at hand. Just because a sickness is rare, it doesn't mean that it should be ignored. Just because an inconvenience is rare, it doesn't mean that it should be dismissed.

Whine and Cheese

A friend of mine told me that when her husband had a hard day at work, she'd greet him at the door with a glass of wine in one hand and a goblet of chips in the other. When one of my husband's colleagues was hospitalized due to MRSA, the deadly, contagious skin infection, I greeted him at the door with a bottle of antibacterial gel. Some men get wine, others get "whine" of a different kind: "Wash your hands, disinfect the computer keyboard and the phones…" The point is, maybe medically speaking, MRSA isn't that unusual in some doctors' practices, but when you're bombarded by news reports and letters from schools concerning its potential dire consequences, it's easy to get afraid. I know many patients who used use antibacterial gel. Some got a rash from doing that, because their skin would get dry. (Sometimes certain lotions can help).

Medical advice can be frustrating, particularly when one doctor's advice contradicts that of another doctor. On the one hand, I want them to all sit around a very large table, and come to some consensus as to what the proper treatment is for a particular disease. On the other hand, I value second opinions, which may or may not be in consensus with each other.

Sometimes, what would be an appropriate treatment for one person, wouldn't be appropriate for another, and may even have negative consequences. Some people are allergic to penicillin, for example, so a sweeping generalization of using penicillin for a certain infection would have a different result on different people. (A warning to doctors and patients alike; be aware of what you're allergic to before you take a drug. This should be a joint effort; a doctor might overlook the information in your chart. The proverb from Luke might be "Physician, heal thyself," but I would add, "Patient, know thyself").

I'm allergic to some nuts, so when I receive a gift of nut-laden cookies for example, I'll always write in my thank-you note, (yes, I'm old school enough to still do that), that my family enjoyed the cookies, because I'm allergic to pecans. Oh nuts!

Just as in allergies, some people have different triggers for different physiological reactions. If someone is claustrophobic, putting them in a small, enclosed room might elicit an unusually negative, strong response, whereas a larger, more open setting might not. Another person might have just the opposite effect. Some people strive on stress; others are overwhelmed by it. I think more and more people understand the need for tailored approaches to medicine, involving not only the medicine per se, but the person who will be taking it.

Car-thartic Experience

I once had to deal with a doctor in a non-medical situation. He was very rude to me and everyone around him, including his family. He scolded me for being nice. He was verbally abusive and derogatory. I was in such shock, that I just stood there and took it. It was indeed the first time that I was "accused" of being nice; usually I'm complimented for that trait.

Anyway, quite some time later, I ran into this man on the road (not literally). He was in the car next to me. I told him everything through my car window that I didn't have the guts to tell him to his face. I was going to have some satisfaction, and perhaps regain some self-respect in the process. I told him he was boorish (not boring, but boorish). I told him he pushed his weight around. I said that in med school, they don't teach you to treat people this way. I reminded him of the Hippocratic (not hypocritical) Oath, that he took when he became a doctor. The oath by

Hippocrates states, among other things, "First, do no harm." What about the harm he was doing not only to me, but everyone around him? Is this the example that he wanted to set, regarding how to treat people? I was on a roll; I felt great.

This must have been brewing and stewing for quite some time. As the lanes merged, I had to take one more look at him. His car was slightly behind mine, and he wanted to get ahead of me. Since I was venting to the window pane, I guess you could call him a "pane in the rear." I wasn't going to ram his car, but I certainly wasn't going to let him in. He didn't want me to be nice anyway, why ruin it for him? And then, on the second look, I realized….that it wasn't him! I didn't care. I still felt great. I waved him in front of me and he waved his thanks. Then I knew for sure that he wasn't the original "real deal," because that person didn't have it as part of his persona to say thanks.

As luck would have it, the light changed quickly, and the man whom I let in front of me got to go through the light, and I was stuck behind the red one. He mouthed, "I'm sorry," into the mirror. If anyone should be mouthing "I'm sorry," it should have been me, because I said all those nasty things about him. I saw him pull into a gas station. If gas weren't so expensive at that time, I would have offered to fill his car up with gas, not only as an apology, but as a thanks, for allowing me to lift a burden off my shoulders. Tanks for thanks, so to speak.

I tell you the above story to advise you not to let a wound fester, physical or emotional. If you have something to say to your doctor, say it politely, close to the time of your concerns. That way, it won't blow up like a volcano. If you just want to vent, as we all need to do, without addressing him or her directly, do what I did. If your doctor's a pain, yell at the window pane. In my case, the "pane in the rear" was the best medicine I could have had.

My normally mild mannered, even-keeled husband was on the phone with an automated voice at a pharmacy, (but at the time, I didn't know that he wasn't speaking to a live person). He must have repeated ten times, "Yes," "No," "Yes" "No". I finally heard him say, "GTH." I was mortified; I told him that he couldn't talk to someone that way, but he assured me that it wasn't a real person. Again, we all need to vent. It's a ca(r)thartic experience.

PDR-Pretty Darn Resourceful

PDR is the " Physicians' Desk Reference." It's a valuable tool providing information on many drugs, including usage, side effects, dosage, etc. I have learned over the years that every doctor can't know everything about every drug. I've known doctors who didn't know that cholesterol-lowering drugs like statins shouldn't, for the most part, be prescribed for pregnant women. I had a friend who had to tell her doctor that with medicine that she was taking, NSAIDS were not appropriate. I say, if you're a patient, do your homework. Your doctor has already done his/her homework to get where he/she is today, but that doesn't mean that they know everything, nor should we expect them to.

The internet is at once a wonderful and a frustrating tool. Reputable websites can provide a wealth of legitimate information. I think it behooves anyone who is prescribing or prescribed a drug, to first look it up on the web, and find out some basic information on it, such as interactions with other medicines, with food, etc. The hard copy of the PDR is a big book; it's not realistic to carry that around with you, but you can carry around some of the information that it provides.

I preface the idea of a patient utilizing the PDR and/or the web for a doctor visit by saying that you should be nice, but be prepared. I wouldn't walk into a doctor's office and demand that he or she read my volumes of information about a specific medicine, procedure, etc. However, I would certainly come armed with information if I had a heads-up as to what the visit would entail. (Heads 'n tails, so to speak). If I were prescribed a drug that was new to me, I would search the web for information on the drug, before I'd do a directional search to the nearest pharmacy, (if circumstances allowed for that).

If you're a patient and you utilize the PDR or similar sources, I say you're my own definition of PDR-Pretty

Darn Resourceful. It can be useful in a "PDR"-
Patient/Doctor Relationship.

Slide Show

I have a relative, a cancer survivor, who was moving to another state. The doctor in her new state requested slides of her cancer to be sent to him, since hers was an unusual form. Much work was done regarding phone calls, request forms, addresses, phone numbers, etc. These types of slides are usually sent via Fed Ex®, from what I'm told, but this time, it slid through the cracks. It went via "regular" USPS mail, never to be seen again. Talk about a slide show. Sometimes things slide through the cracks. In this instance, it was the slide that slid through.

I have gotten tons of mail both important and not, so how this important piece of mail got lost in the shuffle is beyond me. The woman who sent it could describe the packaging in detail, right down to the size of the package. Luckily, there were no ill effects from this, except a lot of worry and aggravation. If it were feasible, my relative would have hand delivered things herself, across state lines, but logistically, it was just more realistic to have this done via the mail. We have to turn our trust over to so many people who are involved in our care: doctors, nurses, mail carriers, etc.

Of course, we have no way of knowing at what point in its journey the slide got way-laid. Was it in the mail, at the receiving institution, etc? It was unfortunately not sent in a way to be tracked. I'm beginning to think that carrier pigeons or Pony Express might be the way to go. There are so many things that are out of our hands, or talons, as the case may be.

I know someone else who had a test administered in one building and was guaranteed that the results would be sent to another building by the date of his follow-up appointment. Not so. It was eventually received, but he was taught a lesson: always double check in a situation like this. Well-meaning, good-intentioned people can give what turns out to be erroneous information, just because of unforeseen circumstances. Therefore, better safe than sorry; do what we were taught to do on tests in school: check your work. In this case, check someone else's.

Sometimes people don't work at 100% capacity. That goes for machines, too. I recently had a routine, screening colonoscopy, and the blood pressure machine didn't work. They couldn't get a reading on me. I thought, "If I've swallowed 8 glasses of stuff in preparation for this, and if I'm already dead, I'm gonna be a little mad."

WMD's—Wonderful MDs

Although in certain political climates WMD's stands for something else, in this book, it stands for wonderful MD's. I have been fortunate enough to have several. Here are some of the qualities that I have valued over the years. They treat you like a person, not just a disease. They don't dismiss your concerns. If they make a mistake, they admit it. (Despite legal concerns, several doctors have done this on several occasions). Admitting their mistakes makes me like them even more. Luckily, none of the mistakes were life threatening. Of course, I'd feel differently if they had been. The WMDs have a sense of humor, and they respect yours. They try to work with you regarding financial concerns, and they don't minimize an extra $50 here and $70 there.

I have only recently found out that March 30 is both " National Doctors' Day" and "Pencil Day". Maybe that's so a doctor can tell an ornery patient to "get the lead out."

On the other hand, maybe a patient would like to tell a doctor that, at times. I guess we'd need a "National Patient Day" for that. Or maybe it should be Impatient Patient Day. For me, I've never wanted to say that to any of my doctors. Most of my doctors have been some of the nicest people whom I know.

My husband and I once ran into one of my favorite doctors in a sandwich shop. After we all left, I told my husband that my doctor looked like he was wearing golf attire, which I thought fitted the stereotype of golf being almost a mandatory pastime for doctors, at which point, my hubby said, "I think I'll buy a sandwich." I told him that we just had lunch, why would he need another sandwich? Of course, he was saying a "sand wedge," (golf club). Well, we had a "long drive" to get that sand wedge, which he used while I "putter"-ed around the house.

The doctor whom we ran into, along with the others, follow up when they say they'll follow up. (I think patients should do the same). These docs don't take themselves too seriously. By this, I don't mean that they laugh while using a rusty needle on you; I mean that they view their profession as just that, a profession, just like countless others. They treat you as an equal. I won't name all of you; you know who you are.

"4 c's" of medical care

We all know the 4 c's of diamonds: "color, clarity, carat (weight) and cut." Shopping for a doctor who'd provide diamond quality care isn't easy. I couldn't care less about a doctor's "color," I do want "clarity" of information, (about procedures, medicine, etc.), I couldn't care less about his or her weight (but I could about my own, remember, "sorry about the weight/wait)," but I do know that if I didn't get along with a doctor, he /she would be "cut" from my list.

I originally thought of 4 c's for medical care, "caring, communication, coordination and cooperation." In reality, there are so many more, but let's address these first. Caring should go without saying. It should be a two-way street between doctor and patient. We are, after all, people first and doctor/patient second.

Doctors- in- training might role play and pretend to be patients. This might give them a sense of what it's like from the perspective of the person in the white paper gown, vs. the white lab coat. This, in turn, might give them more compassion than if they never saw things from that perspective, (including that skimpy gown, chilly room, and sometimes chilly reception). Role playing though, is just that, playing, (pretending) and not the real deal. A real patient experiences fear; fear of the doctor, the diagnosis, the treatment, side effects, the known and the unknown.

Mutual compassion for me is the foundation of any good doctor/patient relationship. If my doctors didn't care about me, they'd be my ex-doctors.

Communication would be next. (These are in no particular order by the way). It's not only communication between doctor and patient, but between patient and office staff, between one doctor's office and another, and even communication within one doctor's office. I once had to have lab work done, and the receptionist told me to come "between 10 and 2." She didn't realize though, (and I certainly didn't) that she had to make a specific appointment within that time frame. This lapse in communication caused a lot of headaches, so to speak. The scenario was repeated over and over, until she got the message.

Speaking about getting the message, sometimes I have left a detailed message with a receptionist about terrible medical complications, only to get a cheery return phone call saying, "Hi, what can we do for you?" Sometimes, the only message a nurse gets is, "Call Mrs. 'Smith' back," with no further details.

I had heard that Oprah Winfrey got many of her doctors together to have a "pow wow" type meeting to discuss her care, so her care wouldn't be segmented, and they could assess all of her symptoms. She's lucky that her doctors do that for her. Sometimes patients are lucky if one doctor faxes another one. We must remember that people are like Aristotle's principle that "the whole is worth more than the sum of its parts."

In addition, if messages don't reach the doctor, how can he/she make decisions about my case? Believe me, when messages have been way-laid, thereby delaying decisions, I've been on the "doctor's case," (even if it took him a while to be on mine). If a doctor doesn't return your call within a reasonable period of time (which can vary from circumstance to circumstance), then call again. I'd rather be thought of as a pest, rather than as a forgotten soul.

I have learned that sometimes mis-communication is due to literally not understanding the person. When I came home from the dentist after having some mild dental work, my mouth was understandably numb. A little while later, I had an issue with another tooth, so I called (with my somewhat garbled speech) to see if my dentist was going to be there the following day and I was told that he wouldn't be there.

The next day I called to see if I could set up an appointment for another day, since my dentist wasn't there then. Lo and behold, he was there. Apparently, when I sounded like I had marbles in my mouth (marbled and garbled speech I guess), they couldn't understand me. Next time, I'll fight "tooth and nail" to find the "root" of the misunderstanding.

Coordination is next on my "wish list." I don't mean physical coordination (although that would be a plus for me). I mean coordination between doctors' offices, MRI facilities, hospitals, etc. Patients shouldn't have to track down records. I feel like I've set the record doing that, and I'm not alone in that feat.

Next would be cooperation, (not co-operations as in two operations at once) although I have had two relatives having surgery simultaneously, but cooperation, not only between offices but within one office. If the office staff is sniping and griping at one another, it often leaves the patient holding the bag, sometimes an empty bag at that. Particularly between doctors and patients, trust is a must, but it can rust if it's a bust. We depend upon everyone's cooperation.

Now we can address the other "c's." My next one would be compatibility, a trait that you can't find in a computer search. Sometimes there's an instant rapport; sometimes it takes time. For me, though, if I'm not comfortable asking questions, then I need to be elsewhere. Of course, I want competence, yet another "c," but I also want the comfort level to ask questions.

I have been lucky; my doctors answer my questions (from the list that I usually bring). Several of my doctors have said, "Good questions." I know people who would never question a doctor, for fear that it would insult him/her to "challenge" his/her judgment. When I taught, students would ask me questions all the time. Frankly, I think it's a sign of intelligence to know what you don't know.

When I was writing out some questions for an upcoming doctor visit, I told my husband that a day without questions from me for a doctor, would be like a day without me as a patient. I told him that there used to be an ad for what I thought was prunes, with the same idea,

"A day without prunes is like a day without sunshine." I told him to just think of me a prune; wrinkled, but good for him.

Well it turns out that the ad wasn't for prunes; it was an ad for the Florida Citrus Commission, and it was for orange juice: "A day without orange juice is like a day without sunshine." He said that was OK too, because I was, to use an old expression, his "main squeeze."

Well, back to the matter at hand. Another "c" for me would be commitment. You need to be committed to following a medical/health plan, (such as eating healthily, exercising, etc.). In addition, you need to be committed to making time to go to the doctor when needed. Caregivers, especially, often put themselves last. Remember to carve time out for yourself. A friend of mine said she wanted to be committed. I knew she was under a lot of stress, and I thought she was talking about being committed to a hospital for psychological counseling. What she meant was, being committed to an exercise program. Can we say "stasis" again?

My next "c" for diamond quality care would be clarity, regarding how and when to take medicine, the purpose therein, the timing, etc. Just as doctors are only human, pharmacists are as well. Once, a pharmacist handed me a drug for another patient with the same last name, but the opposite condition! Vigilance pays off. A relative had a similar encounter, but didn't realize the mistake until after she took the medicine. Luckily, she lived to tell the tale. Some people don't.

The next "c" would be correction, when needed. Medicine sometimes needs to be tweaked, adjusted, etc., especially if you're the "1 out of 100" who might not respond in the usual way. Correction isn't a sign of fallibility, rather, it's a sign of adjusting and adapting, something we need to do in life anyway.

Comfy couches and coffee are yet other "c's" that can make a visit more comfortable, both physically and mentally.

I must add "chuckle" for another "c", something that, as long as it's appropriate, is a welcomed addition to a medical situation. I was once having a mild, out-patient procedure done, but I was slightly medicated for it. The nurse mis-read my name as that of a comedienne, and she said, "Look who we have here, ha,ha, ha." I'm thinking, "I'm the one who's medicated, and she's the one having the good laugh?"

The one "c" that I wouldn't want my doctor to be is condescending. One of the nicest things a doctor did for me was to show me the pharmaceutical insert regarding a new antibiotic that I was going to take. This explanatory leaflet showed chemical bonds, among other things. I wouldn't know a chemical bond from a "James Bond", (the fictional character created by Ian Fleming in 1953) but I do know a nice gesture when I see one. Believe me, when a diagnosis leaves you, like "Bond", "shaken" (if) "not stirred," a nice gesture goes a long way. That's truly diamond quality care.

Another "c" would be conversations. When I had a conversation with a friend about an upcoming procedure, she described it to me as having to lie on a table and turn from side to side for this particular type of X-ray, in which you were injected with dye. She neglected to mention that you are doing this with an instrument the size of a cruise ship inside of you! (Think iceberg). I have learned that she sugar-coated the discomfort involved with this particular procedure.

Towards the beginning of the procedure, my doctor asked me if I was allergic to dye, but I thought he was asking me if I was allergic to dying, and I thought, "Well yes, I'm extremely opposed to it, and would like to avoid it at all costs!"

My mother had come with me to drive me home after this procedure, since my husband was working and it was a difficult day for him to take off. After the procedure was done, my doctor asked me what my mother's name was. I was lucky I could remember my name! Then he asked me what she was wearing, so he could fill her in, while she was in the waiting area. I said that I thought she was wearing a gray top and red slacks, or maybe it was a red top and gray slacks. I told him to find the non-naked woman out there and that would be her.

The beauty of conversations with friends is that you know that they lived to tell the tale, and vice versa. I once told a friend that after a certain procedure, that she'd be in no mood to cook that night. I advised her to cook the night before, and then just stick it in the microwave the next night, or better yet, I told her to have her husband stick it. (You know what I meant. So did she. I still remember our shared laugh).

The beauty of conversations with doctors is that those conversations reveal your fears, pain tolerance (or lack thereof), etc. That's why for me, continuity (yet another "c") with a doctor is so important.

I have to add one more "c": colon, but it's not the type of colon you think. I once overheard my husband say that the Sox (sports team) got Colon, (a player named Colon), but I misinterpreted it as "socks got colon." Socks got colon? What's next, "dress gets pancreas?" (I think the player's name might be pronounced differently, but we didn't know that at the time). Stasis anyone?

Sometimes, just a few adjustments here and there can make a big difference regarding doctor/patient relationships. You'll "c

You are More than Your_____Score

We are so score conscious. We are aware of our credit score, our cholesterol numbers, blood pressure numbers, blood counts and on and on. It never seems to stop. First, there was your SAT score. At the time, you were very aware, now, no one could care. When is the last time an adult asked another adult their SAT score, yet at the time, it was a marker of potential. Now we have biomarkers.

Of course, ranges of appropriate numbers concerning cholesterol, blood pressure, etc, are an important gauge into someone's overall health, but overall, I don't think we need to be obsessed with slight variations with those numbers, unless they indicate a problem. I have also found out that many times those numbers have nothing to do with commitment, attitude or behavior. One person can have fudge but no pudge, and another person can have the reverse. Environment plays a part, but so does heredity.

We shouldn't beat ourselves up if we're not keeping up (or keeping down) with the super slim neighbor next door.

I remember when some tests were graded on a scale, so that if the "best score" was an 85, then all grades were adjusted accordingly. It would be nice if the bathroom scale registered effort as well as poundage. I'm not saying to ignore numbers. I'm saying that you should do the best you can for your overall health, but not to let one annoying but non-life-threatening "score" define you.

I am reminded of a barbecue, where a patron asked the grill master to score the burgers. The man behind the grill, busy tending to the corn, burgers and hot dogs in front of him, hadn't heard that term, and I said, "You know, burgers 2, hot dogs 0." It puts scoring and competition in a more appropriate light. As "corny" as this sounds, I "mustered" the courage to say something funny, and everyone laughed. I "relish" that memory; (lol, laugh out loud), please.

Speaking of keeping score, a friend asked me how many books I hoped to sell, in order for this writing endeavor to be a success. I told her that if one person buys this book, and it makes them laugh, then that's a success. In some respects, it's immeasurable.

Uniform Uniformity

We often think of doctors, nurses and other healthcare professionals as having certain uniforms. I would like to propose that there be a uniform set of requirements from the initial intake of information when a patient enters a hospital, to discharge.

For example, as soon as a patient enters the hospital, and information, often vital, is given to a nurse, all that information should be put at an average eye level on the door. I know of a situation where an intake employee was told that the patient who was brought into the emergency room had a partial upper plate, and that no venous punctures could be given in the right arm, due to the removal of all of the lymph nodes due to a previous lumpectomy. The patient's daughter then walked into the room about ½ an hour later to find, lo and behold, a venous puncture in the right arm!

Apparently, the person who wrote down the information wrote it on a sheet of paper and placed it on a table on the left side of the room. The phlebotomist comes in and goes to the patient's right arm, not looking at the seemingly innocuous piece of paper on the table! And this isn't the first time that this has happened to this patient! Different hospital, same circumstance. History repeats itself. I think that's called, "déjà vu." To add insult to injury, the patient slept with her bridge in, (she could have choked) because no one came to remind her to take it out. Remember, when someone is in a hospital, they are not in familiar settings, so it's all the more important that they are reminded of things that can cause them harm, such as choking on their bridge!

There's "old" in them thar ills. I knew of a hospital that did not use medical bracelets to inform staff of problems, such as medicine allergies, pre-existing conditions such as diabetes, etc. They just asked the patient that information. But an older patient who might be confused about their situation and surroundings anyway might not remember their medicine allergies. It would make sense to have the bracelet as a back-up. (I'm happy to say that this particular hospital has changed their ways).

And for Heaven's sake, look at the information that's posted on the door and/or above the patient's bed! I have had hospital personnel tell me that they didn't even look at that information when blood pressure needed to be taken quickly, even though the sign indicated that no blood pressure cuff was to be used on the patient's right arm. What's the use of alerting hospital personnel, if they don't want to be alerted?

There should be uniform requirements nationwide, so that every person, medical or not, who enters a patient's room, looks in the same place, for information regarding a patient's care. Of course, this only applies to information that is allowed to be shared. I'm all aware of HIPAA laws, but you don't want a patient to break a hippa because of a stupid medical mistake.

Sometimes I feel like doctors and patients' family members are like two (s)hips, or should I say (s)hippas that pass in the night. I can be walking into a family member's hospital room while the doctor is walking out. Just due to the nature of the beast, the doctor is rushing to see someone else, while you're trying to ask questions about or on behalf of your family member. Not that there's an easy answer to this, but again, communication is key.

I would suggest that information be placed in a consistent, uniform place on every patient's hospital door nationwide. You could have a sign on the outside of the patient's door, reminding a sometimes harried doctor or nurse, to look for information on the inside of the door. There, a plastic- type holder could be affixed to the door. The folder holder, if you will, could contain all of the information about the patient that you'd need, including why they are there, medicine allergies, etc. Then that piece of paper would go with the patient. If they were moved from the emergency room to a patient room, for example, that sheet could go into the plastic holder affixed to the inside of the patient's room. That could fix a problem when communication gets lost in the shuffle. When the patient leaves the hospital, they can take that paper with them, and do with it what they want. In that way, if the Electronic Medical Records have a glitch, a hard copy of information can be easily at hand. That doesn't seem so hard to me.

Sometimes, high tech needs to marry low tech, (the way my husband and I did). I think that a PDF computer file stands for "pretty darn frustrating." He knows better. We complement (and compliment) each other. Words to live by, in both a medical setting and outside of one.

They used to say, "What's your (zodiac) sign?" A hospital should be asking, and reinforcing, "do you have a sign?" It could also be placed above the patient's bed. Continual training of new personnel should reinforce not only looking for the aforementioned information, but looking for different procedures and strategies based on that information. "Personalized medicine" is all the rage regarding tailoring medicine doses, types of medicine and chemo treatments, for example, to the individual. At the very least, it should apply to obvious signs. It gives a whole new meaning to the term "vital signs."

There is a prayer from the Old Testament, the translation of which is to place a prayer "on the doorposts of thy house, and upon thy gates, that ye may remember…" I have taken it out of context, and shortened it, to prove a point. Signs should be placed in an obvious spot, so that the appropriate people can be alerted to pertinent information. Philosophers ponder, "If a tree falls in the forest, and no one hears it, did it fall?" I would suggest, "If a sign is there, but no one reads it, what's the good?" Regarding patient information, "you need it, you read it, you heed it." (I plead it).

There's gold in them thar pills.

I had a very wise doctor once who said that anything that crosses your lips, except perhaps for water, can cause harm. This was before a rash of food recalls, but he was talking about medicine at the time. How true. One man's meat is another man's poison. Penicillin can kill one person and help the next. Some medicine that had been recalled was a Godsend to some people, and others got seriously ill, or worse. I think as a nation, we have become used to popping a pill for every ailment without considering the consequences.

Pharmaceutical companies, of course, have to make a living. I just hope that research is done ethically and above board, so that medicine is thoroughly researched, in large studies, to test for possible side effects. Of course, business can have a bottom line, but to the family and friends of loved ones who have died due to taking a pill that might not have been thoroughly researched, money

121

compensation is small consolation. Certainly, some people can have allergic or adverse reactions no matter how carefully things have been researched.

However, there is a difference between an adverse reaction that could not be anticipated, and an adverse reaction due to neglect, improper research, etc. People are vulnerable when they take medicine. The least that we deserve is to know that it has been thoroughly, ethically and honestly evaluated. Thankfully, a system is in place to hopefully ensure our safety.

M.F. Stephenson said or implied that "there was gold in them thar hills," which might have inspired Mark Twain's character Sellers to say the same, but I contend that there is gold (profit) in pills as well.

Of course, sometimes unexpected events happen regarding medicine-unexpectedly negative results, or unexpectedly positive results (think about the discovery of penicillin). That's the intrigue of medical science. Sometimes an off-label use can prove beneficial. There just should be a happy medium between company profit and patient safety.

Winding Up

When I was thinking about how to wind up this book, I was trying to wind down watching an evening news broadcast, (although frankly, it got me wound up). It got me to thinking about how newscasters have ended their broadcasts. Dan Rather used to say, "Courage." Doctors and patients both might need that one; doctors for delivering bad news, and patients for hearing it. Charlie Gibson said, "I hope you have a good day." That's a true sentiment for all concerned.

My favorite, though, is how Katie Couric used to end some of her broadcasts. She'd say, "I'll be here tomorrow, and I hope you will be, too." Although I think she meant it literally, that she'd be at the news desk the following day, and she hoped her viewers would be watching, I think it can be interpreted in the broader sense of the word, too, as in "be here" meaning "being around." I can think of no better way to end a doctor-patient visit than by both the doctor and patient thinking, if not verbalizing, that he/she would be around the next day, and he/she would hope that the other person would be there too.

In fact, I can think of no better way to end this patient/author- reader visit, (patient in more ways than one,) except by saying that I hope that this rather broad broad will be a-round, I mean around tomorrow, in the broader sense, and I hope that you will be too. In the words of my late dog, I hope this book didn't make you snore. In the words of my texting friends, TFR, thanks for reading.